BIG DATA

Analytics

PROJECT
MANAGEMENT

by

Tiffani Crawford

About the Author
Tiffani Crawford designs and builds global Big Data Analytics systems. She has 25 years of high technology experience with Fortune 500 companies, including Cisco Systems, Bank of America, VISA/Inovant, BAE Systems, Applied Competitive Technologies, Ditech Networks, Finance 4 firms, defense contractors and startups. She has worked in seminal technology development in Big Data, analytics, cloud, networking, telecommunications, software development, distributed multi-tier applications, multimedia/digital, geographic information systems, intelligent transport systems, finance, data modeling, security, policy systems and structural equation modeling. She is a credited software developer and published author. She earned her PhD from the University of Southern California in 2005. She has also earned her Master's, Bachelor's and various technology certifications. She is a member of PMI with various philanthropic contributions.

Tiffani Crawford is available to answer questions about Big Data Analytics Project Management and related topics via email at zapquick@aol.com.

Many thanks to all of the people
who contributed to my wisdom.

Contents

1

Introduction to Big Data Analytics

Big Data Analytics technologies offer substantial benefits. When we evaluate these technologies, we have many questions.

Is there a business case for Big Data Analytics?
- What is Big Data Analytics?
- Getting to business value
- Opportunities and implications
- Clients

How does Project Management for Big Data Analytics bring in the business value?
- Common misconceptions
- Do Big Data Analytics fit your client's interest, situation and experience?
- Choosing the best methodology
- Evaluating caveats and risks
- Unique considerations
- Growth and sustainability

Like project management, Big Data Analytics require expertise to make projects successful.

Big Data Analytics technologies require a breadth of ecosystem tools. Big Data provides the right type of data processing for the data. Analytics query the data to gain insights and visualize them for people's understanding.

An effective Big Data Analytics system includes a data processing layer and an analytics layer. If there is no business value from the data, there is no reason to process it and query it for meaning.

Business Value = VVCV+SQVV+VVVV

Big Data Analytics technologies are advancing rapidly. As we learn more, we grow more data and develop more wisdom about using it.

When people mention the 3 Vs, the 4 Vs or VVCV, they're talking about the nature of Big Data. Understanding Big Data characteristics helps us qualify the client's data problems as good ones for a Big Data Analytics system to mitigate.

"Big Data is coming in fast – are we ready for it?" Clients often do not know how to ask themselves the question or find the answer.

The evolving formula BV = VVCV+SQVV+VVVV can become very technical when we discuss designs.

Business stakeholders cannot get through these details, preferring to focus only on the outcome: Business Value. Without business value, there is no point in using Big Data Analytics.

For the business stakeholder, we begin with large quantities of many data types: structured, semi-structured, unstructured, human and machine.
- Structured data is defined data: email = address@domain.com
- Semi-structured data requires some format
- Unstructured data is freeform, such as blogs
- Human data comes from people's actions
- Machine data comes from the interactions of computers, mobiles and servers

To produce the business value, we must apply analytics to the data we've captured and processed in our Big Data system. Analytics give us insights when analytical methods and models are used.
- Quantitative, or numerical
- Qualitative, or non-numerical
- Correlation, or Pearson
- Longitudinal, or time-series
- Social, from social media
- Search, from semantic search engines
- Operational, from platforms and machines
- Inferential, from extrapolation of known data
- Ethnographic, for people, identity and history
- Interview-Based, from experimental interaction
- Casual, from randomly collected data

Intelligent analytics are based on valuable data points among the captured data. The system can perform counts, comparisons and calculations in accordance with applicable conditions and rules.

Analytics algorithms sift through large amounts of data like an astronomer scanning the universe. One data point is a planet in our solar system. Another data point is a quasar in another galaxy. That black hole only matters if the customer is aged 18-24. If both suns are present, divide by 8.

When should we use Big Data Analytics?

We must analyze VVCV+SQVV+VVVV to determine if our problem is best mitigated by Big Data Analytics. If one or more of the following data characteristics are present, the client can benefit from Big Data Analytics in the network for the product and corporate systems and platforms.

Volume

Big Data is high-volume.
- Transactions per Second (TPS) in Terabytes (TB)
- Storage per Day or Year in Petabytes (PB) or Exabytes (EB)
- Hadoop nodes = 30+ planned
- Data source systems = 10+ planned
- Dedicated nodes for processing, analytics, data streaming or security
- Multiple data hubs
- Multiple datacenters

Variety

Big Data leverages many types of data.
- Structured
- Semi-Structured
- Unstructured
- Real-time data streaming
- Mobile data
- Machine data
- Imaging data
- External data
- Metadata

Complexity

The complete journey from data intake to business value is not easy. Big Data leverages complex relationships among other data, applications, infrastructure and people. Analytics requirements include processes and related algorithms.
- Predictive statistics anticipate what will happen, an advancement over knowing only what has happened.
- Data architecture is specialized.
- Data pipeline analysis is needed.
- Computational models include self-learning.
- Insight granularity is customized and specific.
- Query and results are packaged for usability and interoperability for analytics applications.
- Application process completions are monitored and audited for compliance purposes.
- Extensive metadata is leveraged to describe content in data records and related files.

Velocity

Big Data comes at high speed in real time. Analysts query and expect results fast.

- Data streaming
- Data processing in Real Time (<5 minutes) or Near-Real Time (NRT) (<30 minutes)
- Query and results in <10 seconds
- Visualization streaming
- Business reporting
- Application processes and networks

Security

We must provide any necessary data privacy and security through our data models, data masking, database encryption, firewalls and other means. We have not achieved data quality if someone's privacy and security have been compromised.

Quality

Big Data requires clear data architecture and data governance to produce data that can be validated easily as it streams in from diverse verified data sources. This produces the data quality for which we can then get business value through analytics.

Verification

During data intake, we must determine that the incoming data is from the correct person, service and locality, for example, through GPS and IP address lookup.

Validation

We must also use the most current data values and remove old data and unneeded copies. Text handling may be needed to resolve misspellings and other user data entry errors along with data uptake errors from application bugs and transmission conditions.

Veracity and Data Story Truth

Big Data must be correct in its indicative meaning in order to provide business value through Analytics. Data must tell a true story.

Data in itself is not necessarily truth. A data point can be erroneous or incomplete.

A data point may not be indicative and may not be able to tell a story. For example, a person's GPS at a café may not indicate that she will buy a new mobile from the shop next door in 20 minutes.

An integrated view across multiple data sources can cross-check for data story truth and discover indicative meaning as part of predictive analytics.

A customer has story truth when following an interest or seeking a product or service. In Scrum, a requirement has story truth. Among various data from diverse sources, an analytics requirement has story truth.

The only way that we can know data's true story is through data processing and analytics.

We may produce a 360-degree customer view for advanced personalization of the customer experience, or we may keep our network provisioning a step ahead of our customers' spiky usage. For the benefit of people and technology, data story truth matters.

Variability

Some data values and types are static. For example, a person's userID is always a string. The person cannot change the userID after registering for the online account.

Some data values can be changed. The nickname, screen name, avatar, image and other elements can be updated by the user.

Even data types can be changed. If you prefer an avatar (image) instead of a nickname (string), you can change it in your profile. Some network data types can also be changed.

Values in a range, such as the service bars available for mobile calling, vary with some observable pattern, including deviations from that pattern. Software can monitor adaptive values and take action based on rules.

Groups of data types and values can follow patterns. The patterns can be analyzed for deviations consistent with security issues, adverse weather events, negative comments indicating product problems, and much more. Analytics can leverage pattern adherence and deviation data for compliance reporting to demonstrate incident prevention and variation control.

Virtualization

Big Data Analytics systems can be virtualized like other systems with a grouping of virtual machines managing the nodes. Network management, provisioning, configuration management, release automation, test automation and more can be virtualized. In addition to more complex tools designed for production, open source tools are designed to install quickly on a small pilot cluster to facilitate initial configuration, traffic analysis and load balancing.

Vexation

How troublesome is it to implement and maintain the system that is really needed? What about fulfilling on Service Level Agreements that require 5-9s (99.999%) uptime and data accuracy with all-or-nothing transactions completing in real time? Big Data processing systems are inherently resilient, reducing vexation, downtime and support calls. Big Data Analytics can mean the difference between business success and costly failure.

Big Data Analytics Initiative Challenges

A Big Data Analytics initiative requires a serious commitment to transform thinking and ways of working along with technology.

People

Specialized skills are required and hard to find. Skills must bridge areas of deeper expertise, but the skills do not reside in the same people.

Application development is more exploratory than quick-churn coding. Overaggressive task decomposition can leave critical impacts on the application and infrastructure undiscovered. The attempt to control cost through tasks can produce much larger costs in production.

Application, database and infrastructure administration is often decentralized along with knowledge about the systems in play. Security and expert access requirements are granular.

Solution architecture is a required perspective.

Business analysis is a necessary input focused on the business value and how it is measured.

Data science, an emerging field, is difficult to staff. Data analysts can groom data for data scientists and their deeper expertise.

Project management is emerging for Big Data Analytics efforts and systems. Expert project, program and portfolio managers are rare.

Executive support and sponsorship is rarely expert even among IT stakeholders.

Data Quality and Management

The enterprise must establish data architecture in order to enable data validation, quality and governance. Without clear data architecture, analytics insights cannot be gleaned and business value cannot be delivered.

Metrics and KPIs must provide an initial analysis and have growing room. Just like there are always more data sources, data types and use cases for Big Data Analytics, there are more measurements and refinements to them.

Several Big Data Analytics project models include data architecture, validation, quality, governance, management and metrics/KPIs. These are required for any Big Data Analytics effort, but some project models assume that significant analysis and building will be needed: Big Data Hub, Big Data Factory, Big Analytics Factory, Big Data Foundry, Big Data Lakes, Big Data Center of Excellence, Big Data Analytics Delivery PMO, Big Data-as-a-Service and Big Analytics-as-a-Service.

Data Centralization

Centralized Big Data quality and governance must facilitate and enforce business unit compliance, making it easier for BUs to use Big Data Analytics.

Data Decentralization

After centralization, some BU activities can be decentralized in proximity to the usefulness of the data and the people working with it. A mature Big Data Analytics system implementing use cases for business units can adapt interface requirements as needed for special data.

Raw data and analytics results can be made available company-wide if desired.

Facilitating tools are critical to the success of the Big Data Analytics initiative. Expert users need their specialized applications to interoperate with the Big Data Analytics system. Many plug-ins, frameworks, libraries, dictionaries and other tools exist to make Big Data Analytics easy to adopt and integrate.

For the Big Data Analytics initiative to succeed, the system, integrations and tools must be usable and offer valuable data visualizations to a variety of users for different purposes.

Because of the amount of data handled by a Big Data Analytics system, business units may be concerned about network performance, especially if charge-backs are in place to penalize high-traffic groups. Operating costs and benefits must be reviewed holistically so that the full value of the system is understood easily.

Technology

The application architecture must distribute functions across nodes with resilient load balancing and traffic management.

The infrastructure and network must fully support a complete Big Data Analytics ecosystem.

Training and support must be provided for a successful adoption of Big Data Analytics technologies.

Change Management

Transformational change management must drive business value. The Big Data Analytics initiative has an information focus.

Big Data Analytics Thinking is part of the client's transformation from less effective ways of doing business internally and externally.

Transformational change management analysis can help the client get the most from the initiative. Some planning must involve company culture and reward systems.

Planning

Information Lifecycle Management is a critical aspect of planning. In a Big Data Analytics system, data is alive and useful in real time. Along with data processing, analytics insights can occur in near-real time. Most considerations involve how to get the most out of the data now. This may require replanning as new use cases are requested after the initial rollouts.

The balance of Big Data Analytics centralization and decentralization in the enterprise requires planning with a degree of foresight. Specific responsibilities may vary with stakeholders as more business units join the initiative.

The traditional considerations of storage durations, locations, policies, privacy and security are still important and must be addressed as well.

Technology-to-Business Value Process

Getting from Technology to Business Value is a process with specific activities in sequence to deliver optimal business value.

Step 1 – Consume static and dynamic data from multiple distributed sources. Analyze first and then store the data.

Multiple data types and sources are quickly fused and analyzed. Analytics in-stream and in temporary memory – before data goes to long-term storage – save infrastructure costs by 10x to 100x, reducing data storage and transport.

Step 2 – Leverage data fusion, streaming analytics and intelligent caching to analyze huge data volumes in real time. A system typically deals with over 250 billion transactions per day and over 100 terabytes of data per day.

Step 3 – Provide actionable, timely insights by visualizing the meaningful business context for key functional needs such as marketing analytics, customer care and network operations. Focus on functional business needs with an eye open for reusability when possible. Dynamic, interactive applications in near-real time drive value.

The Business Case for Big Data Analytics

Big Data Analytics are applicable to all enterprise applications and clients, providing many benefits. Faster, cheaper, easier and more reliable systems can benefit any product or corporate function.

Big Data Analytics systems increase profit and reduce risk. Predictive analytics become possible with proper handling of the data that is really present in information systems.

Faster
- High-speed data processing
  - 1 big job => many small jobs (reduce complexity)
  - Redundancies prevent data loss.
  - Temporary storage is faster.
  - Long-term storage is easier to access.
- High-speed query-results processing (<10 sec)
- High-speed predictive analytics and responsive models

Cheaper
- Low cost (2-10x cheaper) up to massive scale
- Technology-agnostic
- Commodity hardware
- In-Memory options
- Interoperates easily with plug-ins
- Open source available

Easier and More Reliable
- Data validation and quality = near-100% accuracy
- Network resilience and 5-9s performance
- Security resilience and management
- Virtualization and automation
- Network traffic management
- Transitory data management
- Data retention policies and management
- Enterprise-wide analytics with quality data
- Query management and reusable code

Start Small and Practice
Big Data Analytics must be implemented in a structured way.

Begin by defining the initial project scope and business value:
- A good-sized bite of functionality
- An analytics question
- A specific use case
- A necessary measurement

When the project can be stated as a use case, define it and its metrics/KPIs. How will business value be measured?

Evaluate options to leverage open source, public or private cloud, and/or a large supplier install base. On which servers will the project run?

Is this small project a Proof of Concept, Proof of Value, or Proof of Technology? The goal impacts the results.
- Proof of Concept (POC) tests the desired functionality.
- Proof of Value (POV) determines how the desired business value can be delivered by the tested functionality.
- Proof of Technology (POT) verifies that the technology works as intended.

Is early analytics modeling or a Proof of Model (POM) needed?

Even though this is just a small project, it is wise to plan for the pilot, testing, pre-production and production rollouts.
- Is this project going to be tested in staging or production?
- Is this project going to become part of a pilot?
- When and how will the project go to production?
- What happens to the infrastructure used in the project? Can it be reused as in the QA, staging, pre-production or production environment?

Who uses Big Data Analytics?

Large, medium and even small enterprises use Big Data Analytics, including:
- Adobe
- AOL
- Bank of America
- eBay
- Facebook
- Google
- Hulu
- Information Sciences Institute
- LinkedIn
- Quantcast
- Twitter
- WildPackets
- Yahoo

2

The Ecosystem

Technology-Agnostic Market

The Big Data Analytics market has defined itself as next-generation data warehousing and business intelligence, with players providing basic components and industry-specific solutions including products and services.

The most business activity is focused on Research & Development, entrepreneurship, business transformation and market extension. This brings in stakeholders from across the enterprise. Big Data Analytics do not just concern IT and IS departments. Data insights and systems are useful whenever predictive analytics can make a difference, including corporate functions, product development, marketing and sales.

The market is doing well with the basic four Big Data characteristics, VVCV:
- Volume
- Variety
- Complexity
- Velocity

The rest of the evolving formula, SQVV+VVVV, is still emerging as a set of defined solutions for virtualization, configuration management, ecosystem monitoring and software-defined networking.

The opportunities for service offerings are rich.
- Consulting
- Data governance
- Solution engineering
- Research & Development

The execution components have developed into their own niches and are poised to differentiate, integrate and converge.
- Business context
- Information delivery
- Analytics and insights
- Processing
- Integration and management

Companies providing a package of products and services as an industry-specific solution are likely to gain the most revenue from their Big Data Analytics efforts.

Fortunately, Big Data Analytics is available as both open source and proprietary code running on commodity gear. Applications can be built or purchased off-the-shelf.

The Big Data Analytics ecosystem includes these players and applications. For more information, search for them online.

- AWS
- Apache
- AsterData
- Avro
- Chukwa
- Cisco
- Clarabridge
- Cloudera
- DynamoDB
- EMC Greenplum
- Flume
- Ganglia
- Glassfish
- Google
- Hadoop
- Hive
- HornetQ
- Hortonworks
- HP HAVEn
- IBM Big Insights
- Informatica
- Intel
- Java
- Kerberos
- Linux
- Mahout
- MapR
- Microsoft
- MicroStrategy
- Nagios
- NetApp FlexPod
- Netezza
- Oracle
- Pig
- Python
- RabbitMQ
- Red Hat
- SAP HANA
- SAS
- Scribe
- Solr
- Splunk
- SPSS
- Sqoop
- Syncsort
- Talend
- Teradata
- Vertica
- VMware
- Wackamole
- Zettaset
- Zookeeper

The 4 Pillars of Big Data Analytics

Hadoop
- Java or alternatives
- SQL
- Components in ecosystems

NoSQL
- Alternative to SQL databases
- Components in ecosystems

Hybrid Systems
- Hadoop and/or NoSQL components
- Predictive analytics
- Legacy databases and Enterprise Data Warehouses (EDWs)
- Legacy analytics for understanding what happened
- Install bases from large providers
- Slow ETLs moving data

Data Science
- Analytics for prediction and business transformation
- Data visualization and reporting

Hadoop

Sometimes people talk about Hadoop as if it were an entire system. Hadoop is data processing with limited analytics written in Java code language or an alternative such as Python used with an adapter. Hadoop is SQL-based.

Hadoop's two components are MapReduce for data processing and HDFS for data storage.

MapReduce takes a large data document and breaks it into many small data documents for faster processing on redundant nodes. At the end of the process, the one true dataset remains for analytics and storage.

The Hadoop Distributed File System (HDFS) is the database where the processed data is stored.

Note that the ecosystem is not complete with only Hadoop. Much more is needed to fully enable and secure the system's applications and infrastructure.

Typical Hadoop ecosystem components include MapReduce, HDFS, Hive, HBase, Scribe, Zookeeper, Avro, Chukwa, Sqoop and Pig.

NoSQL

SQL code can be messy to process and store, so the NoSQL alternative was written.

NoSQL is not relational algebra feeding RDBMS tables. The data manipulation language is not SQL.

Most applications and databases today can handle both SQL and NoSQL using plug-ins and adapters. It is possible to use both SQL and NoSQL components in the Big Data Analytics ecosystem.

There are many NoSQL options for various components in the ecosystem.

- Cassandra
- Couchbase
- CouchDB
- Google Bigtable
- HBase
- Hypertable
- MongoDB
- Project Voldemort
- Redis
- Riak
- Scalaris
- SimpleDB
- Terrastore
- Tokyo Tyrant

Hybrid Systems

Most clients have an Enterprise Data Warehouse (EDW) with related Data Marts (DMs) and Operational Data Store (ODS) in place today. Many have business intelligence and data visualization applications along with existing integration for data from enterprise systems and external sources such as social media.

Because of the prevalence of hybrid systems in the market, some applications deal well with this.

Traditional Enterprise Data Sources
- IBM DB2
- Microsoft SQL Server
- Oracle
- SAP
- Siebel

Typical Social Media Data Sources
- Facebook
- Geodata (GPS)
- LinkedIn
- Mobiles
- Twitter
- YouTube

Data Integration
- AbInitio
- IBM DataStage and InfoSphere
- Informatica
- Pentaho
- SAS
- Talend

Business Intelligence and Data Visualization
- BusinessObjects
- Cognos
- Datameer
- Information Builders
- Jaspersoft
- Microsoft
- Oracle OBIEE+

- QlikView
- SAS
- Tableau

Data Science

Making sense of data is both Art and Science. With well-envisioned and structured data science, enterprises can:
- Generate better insights
- Gain confidence in decisions
- Visualize the data
- Understand the data and communicate that
- Learn how to learn and adapt with agility

Methods

Data science methods are applied discretely unless the data question is best answered with a multi-method approach.

Data mining sifts through volumes of data to aggregate and relate various data points.

Machine learning applies adaptive rules to defined datasets to provision and govern machines.

Artificial intelligence leverages adaptive rules so that more refined rules can be used automatically.

Information retrieval simply pulls data as queried.

Statistical analysis uses statistical methods to analyze data for insights based on correlations, ANOVA, time-series and other analyses.

Gap analysis seeks to find the missing data or growth path between the current state and the desired state.

Programming Languages
The 2013 Top 10 Analytics Programming Languages were determined by a yearly survey of data science workers. Half of them and the top 3 are all used with Big Data Analytics.
- **R**
- **Python**
- **SQL**
- SAS
- **Java**
- MATLAB
- High-level data mining suite
- UNIX shell/awk/sed
- C/C++
- **Pig Latin, Hive and other Hadoop-based languages**

For more information, search on these languages.

Applications

Analytics and data visualization applications can be general to the enterprise or specific to a type of industry-based solution.

- BusinessObjects
- Cambridge Semantics
- Clarabridge
- Cognos
- Datameer
- DOMO
- Eagle Genomics
- Google Bigquery
- IBM SPSS
- Informatica
- Information Builders
- Jaspersoft
- Mahout
- Microsoft
- MicroStrategy
- Netezza
- Oracle
- Pentaho
- QlikView
- R
- SAS
- Semantelli
- Syncsort
- Tableau
- Tibco Spotfire

Typical Visualizations

The most common visualizations are available in nearly all analytics applications.

Heat Map – On a map, circle size and color indicate the amount of comments and whether they are positive, neutral or negative.

Klout Score – Rankings by the amount of discussion based on who speaks and who listens or replies.

Flow Map – The amount of data flowing from one datacenter to another is indicated by the weight of the line.

Rotating Globe – Normally flat visualizations of Heat Maps and Flow Maps can be plotted to display on a rotating globe. Clients are engaged by this display.

Link Analysis – The Heat Map pie charts show relative size and sources of revenues in correlation with traditional bar chart displays.

Bar Chart in Real Time – New customers top existing ones on a bar chart updated in real time.

Service Provider Essentials – On a map, the service provider's office is noted with name, address, distance, contact options and wait time for customers to be served.

Influence and Spending – Charts correlate professional conferences to professional purchasing.

Ecosystem Architecture

When building the Big Data Analytics ecosystem, think in layers. It is easier to start small and grow. POCs, POVs, POTs, POMs and pilots can be tested and implemented much faster with clearer business value as the system is built.

These layers form the best practice of building in repeatable steps with reusable code.

Transaction Data is easy to hook up first with a simulator before taking test account feeds.
- Online Transaction Processing (OLTP)
- Online Analytical Process and Data Warehouse Appliances (OLAP)

Interaction Data can also be simulated, but testing works best with test account data. Sometimes new data types and fields are present that were not part of the UI days earlier.
- Social Media
- Clickstream
- Text
- Audio, Video
- Mobile, CDR, GPS
- Machine, Device
- Scientific, Sensors
- RFID

Big Data Integration involves fused data querying.
- Mediated Querying
- Portals
- EDW Data Warehouses
- Operational Data Stores
- Federated Database Systems
- Workflow Management Systems
- Peer-to-Peer Integration
- Personal Data Integration

Big Data Processing manages data through the reconciliation and deduplication processes. Analytics can be run on data in temporary storage.

- Selective query
- Petabyte sorts
- Checksum across nodes
- Navigational search
- Text mining

Big Data Storage packages datasets within HTML wrappers for easy query and retrieval.

- Query availability
- Fault tolerance
- Load distribution
- Coherent execution

External Recipients exist outside the enterprise firewalls for data results based on queries and data pushes based on rules.

- Data Hubs and Aggregators
- Compliance and eDiscovery

Big Data Analytics provide a variety of benefits. Predictive analytics capability is by far the most important in the enterprise today. Reporting what happened is not enough for companies to succeed at the speed of real global business today.

Along with their timeliness, predictive analytics make it possible to monitor, predict and prevent negative events and harmful social conversations from happening.

Most predictive analytics are used to increase revenues by better understanding the customer, better anticipating customer needs and responding with offers and products, and lowering the cost of compliance, risk management and revenue generation.

- Community-building
- Market basket analysis
- Social analytics solutions
- Micro- to Nano-level customer segmentation for better financial service targeting
- Regulatory and risk compliance management
- Innovative risk management-based business models
- Predictive revenue and risk modeling for longer time periods

Think of this complexity in simple steps to build from pilot to production.

- Data acquisition and querying
- Data ingestion and verification
- Data reconciliation and deduplication
- Data storage and query results packaging
- Analytics preprocessing
- Analytics data mining
- Analytics storage and results packaging
- Failover and disaster recovery

3

Adopting Big Data Analytics

The adoption of Big Data Analytics technologies follows 4 distinct phases for most enterprises. Each phase includes activities that are critical to reaching the next phase.

Learn – Start small and gain experience
- Technology evaluation
- Basic infrastructure
- Technology Proofs of Concept
- Knowledge and skills

Implement – Test and build out
- Reports and analyses from Big Data
- Big Data Store established
- Social and sentiment analysis enabled
- First set of business data integrated
- Business POCs and pilots
- Technology vendor ecosystem
- Big Data focus group
- Big Data cluster established

Optimize – Make the most of data
  − Pattern analysis and discovery
  − Comprehensive data platform for data scientists
  − Business analytics on structured data enabled
  − Text parsing and analysis
  − Machine learning for pattern discovery
  − Unstructured data integrated
  − Big Data roadmap and engagement model
  − Big Data architecture and design standards

Innovate – Go predictive!
  − Big Data Analytics-driven business innovation
  − Business strategy modeling
  − Deep data mining
  − Predictive modeling and decision insight

Vendor Selection

Even though Big Data Analytics involves new technologies, traditional data warehouse and business intelligence vendors are adapting their stacks to include Big Data Analytics. The most common model is to run the application, via Cloudera, Hortonworks or MapR, on top of an optimized hardware stack. This is the model in use with Cisco, HP, IBM, Intel, Microsoft, NetApp, Oracle, SAP and Teradata Aster.

If it is not possible to put in a large-scale install base, an enterprise can begin with the implementation of better components on various platforms. Data offloading from slower, costlier systems is a popular approach.

Another approach is to develop custom applications and components in-house or through a development services and staffing vendor. Some functional areas, such as development, testing and 24x7 support, can be provided through a vendor as well.

For help shortlisting vendors, review industry reports from popular online sources such as Forrester and Gartner.

Top 7 Enterprise Hadoop Solutions, Forrester 2012:
- AWS
- IBM
- Greenplum
- MapR
- Cloudera
- Hortonworks
- Pentaho

Top 7 Data Warehouses + Hadoop, Gartner 2013:
- Teradata
- IBM
- Oracle
- SAP
- Microsoft
- EMC
- Actian

Top 7 BI + Hadoop, Gartner 2013:
- IBM
- Microsoft
- SAS
- Oracle
- MicroStrategy
- QlikView
- Tibco Spotfire

Opportunities and Implications

The purpose of Big Data Analytics is business value delivery. When we enable an optimized system fully for predictive modeling and decision insight, we create many opportunities for business growth along with implications that require some thinking.

Research & Development Pipelines

Research process automation, analysis and discovery improve by scales of magnitude when Big Data Analytics are in place. Various complex technical systems such as genomics, semantics, energy management, global infrastructure and entertainment animation pipelines become easier to sustain and enhance. Integration with social media adds the experiences of real people.

Intellectual Property Protection

Intellectual property moves from the patent protection arena to a fully-enabled predictive brand protection system including social media and compliance feeds using Big Data Analytics.

Mission-Critical, Always Available

Big Data Analytics systems provide redundancy, reliability and resilience by design and capability. Mission-critical systems such as health care, emergency response, national defense and law enforcement require 5-9s (99.999%) availability for data querying and results, sometimes on a global basis. For example, an abducted child may have only hours or days of life. Systems that catch child abductors must work quickly and accurately every time.

These systems require 5-9s accuracy for data results upon high-speed queries, for which the best query-results design involves an all-or-nothing transaction processing model and a refinement to the classic ACID vs. BASE question. The data provided by these systems must also stand up to the scrutiny of criminal trial proceedings and the protection of individual rights.

Reduced Cost and Rework

Most clients want to reduce cost by using commodity hardware, automating manual work, and reusing code, frameworks, tools and services whenever possible. Big Data Analytics ecosystems reduce the costs of applications, infrastructure and people, freeing resources for more complex, valuable work.

Once data insights are proven valuable, the idea gathers momentum across the enterprise. Cost reductions can easily be applied to advancements.

Big Data Analytics Strategy and Roadmap

A Big Data Analytics strategy and roadmap are required to manage growth. Implementations can cover a well-defined portion of the enterprise network. The Enterprise Data Warehouse (EDW) and Operational Data Store (ODS) are good starting points.

A cohesive strategy and roadmap can prevent haphazard implementations. For example, a cost-cutting initiative may prompt a variety of data offloading projects to phase out the more expensive large vendor stacks in favor of commodity hardware and open source applications leveraging Big Data Analytics. A planned progression of implementations allows the enterprise to start small, learn and reuse.

Collaborative Leadership

The expertise to implement a Big Data Analytics system resides in a variety of roles in the enterprise and on the project. Teams must collaborate at all levels of granularity and rollup. Requirements must be validated collaboratively and conveyed to anyone working on a related part of the system. The roadmap and PBI must be visible to the team.

For example, security requirements can become confused in the development of a new system:

Product Owner: "This database must be fast. Hadoop data processing is implemented behind the firewall. We do not need database encryption."

Architect: "We must turn on encryption in the database, or this system will not pass CTO and CSO approval. One publicized security incident can kill this product and possibly the company."

Project Manager: "We will complete data modeling under the assumption that we will start with unencrypted data, clean up functional issues, and then update the model for data encryption. We know the system will be slower, so we will optimize the processing speed for queries and results."

ScrumMaster: "We will take up this requirement at the right time."

Product Owner: "This requirement is for unencrypted data."

Scrum Team: "We'll build for unencrypted data."

Developer: "Here is the code."

Network Engineer: "Here is the VM."

Database Administrator: "The database is implemented."

QA Tester: "The database works. Here are the documented test results."

CTO and CSO: "Why isn't this encrypted?"

DBA: "This code is for unencrypted data."

Developer: "This requirement is for unencrypted data."

Product Owner: "The customer will not accept a slow system."

Project Manager: "Wait a minute. We're not done yet. Now we must update the system for encrypted data and retest. Then we will optimize processing speed. With this experience under our belts, we can begin with encrypted data going forward as we bring in new data types."

Expert requirements analysis can help determine which requirements are complex when first accomplished and then become easier, more repeatable and even automatable later in the project. Some requirements take two or three involved steps and thus need two or three sprints. Take the steps needed to move forward logically and with repeatability in mind.

The Learning Curve

The learning curve may be steep for some or all of the team members on a Big Data Analytics project. The team may protest, "But we've never done this before! We can't guarantee the scope or schedule. We don't know what we don't know."

This is true. Do not panic.

Teams collaborate best when they learn among peers. Remember to encourage experimentation. Do not fear failure.

Big Data Analytics thinking requires transformation in the enterprise. Project managers must encourage calm collaboration and enjoyment of the activities.

Big Data Analytics thinking is new and offers great learning opportunities. Be sure to support formal learning for team members. Plenty of free training and documentation can be found online.

There are also vendor courses, boot camps and certification programs to help teams adopt Big Data Analytics. Customized training from a vendor may be the best option if many teams and stakeholders are involved or we wish to attract their participation. Clients demand education programs and technical workshops to help them figure out what to do with Big Data Analytics.

Rewards and Fairness

The hype and attention around Big Data Analytics can compel more budget for projects, raises for team members, and many concerns in the rest of the enterprise. As the Big Data Analytics team is acknowledged as driving more of the company's revenues, internal politics and interpersonal jealousies can occur.

Big Data Analytics is a spearhead of business value for the company. The new technology will one day be ubiquitous, a part of every data system in the enterprise. As the need for Big Data Analytics grows, the core team can involve other teams, spreading around the accomplishment and raising the water level for all boats in the company.

A balance of vertical and horizontal equity is the most effective in enterprise financial policies and in the assumptions made about the meaning of data about people. Vertical equity means that the person who did the hard work is the one who gets the rewards of that work. Horizontal equity concerns a minimum standard of living, rights and participation for everyone.

When we leverage horizontal equity to preserve work efforts that are not large revenue-generators or cost-preventers, therefore balancing the direct rewarding of vertical equity, we reduce some of the risks of failure. This principle is integral to

Earned Value Management (EVM) and Life Cycle Management (LCM) as well as general democracy.

There's a lesson in this for data interpretation as we balance data privacy needs in the use of the data collected. As we capture more data and analyze it predictively, we learn more about how to optimize our economic systems. For example, fiscal policy analyses such as the Greater Tax Base (GTB) model compute equity based on the value of the dollar as a portion of the earner's income, showing how to simultaneously reduce the tax amount for everyone, limit the tax burden on the middle class and prevent regressive taxation from impacting the lower and middle classes. As we continue to deploy pattern-based data analytics, we must remember that other patterns and sometimes the lack of a discernible pattern characterize the economic drivers of entrepreneurship and disruptive invention. As individuals, teams and enterprises, we must continue to grow our capability and remain engaged in the business value of what we do.

Clients and Projects

Clients may say that they don't know what to do with Big Data Analytics, but stakeholders express clear needs and interests. Part of this is in response to hype and part is an extension of traditional thinking about data warehousing and business intelligence systems.

These are the top efforts that have garnered clients' interest in Big Data Analytics in 2013:
- Enterprise data management, architecture, governance and validation
- Financial data management
- 360-degree customer views
- Business development analytics and decision-making
- Operational data management
- Data query-results systems with fast processing power
- Investment data validation
- Social media, people-interest matching and innovation discussions

Trends and Future

Big Data Analytics trends show at least some of what the future holds for these technologies.
- Renew – New thinking revitalizes companies.
- Adapt – Existing models and open source applications are proven.
- Innovate – Advancements are facilitated.
- Streamline – Efficiencies can be leveraged.
- Extend – Carry the business value forward.
- Automate – Reduce vexation and focus on more valuable work.
- Consolidate – The industry has many players poised for M&A activity while technologies in their first 5-10 years can be leveraged together.
- Discover – There is always more to learn.

4

Project Management

Where does Project Management fit in?

Hadoop was used first in 2004 and became better known by 2007. It is 9 years old. Originally, open source Hadoop promised that projects did not need project management because it was so easy to get the system together, up and running.

Today, some clients process Petabytes of data per day globally. Transactions reach Terabytes per second. The definition of Big Data processing has expanded to include more technologies. Big Data Analytics form a complete ecosystem.

Good open source is still available for enterprises to build and enhance systems through a series of small projects. A large platform takes expertise and time to build, involving 50-100 engineers and requiring project management.

Large infrastructure providers offer complex, expensive systems for vertical clients. These prebuilt systems require customization, planning, staffing and full project management.

Big Data Analytics Project Management is required for project success. Project managers must have technical expertise, business sense, and expert judgment in scaling and working with the technologies. Program managers must synergize the efforts across projects, facilitate as needed and proact in a changing environment.

Experts are hard to find in the job market.
- Hadoop Project and Program Managers – 1 expert for every 4 open positions
- Hadoop Developers and Administrators – 1 expert for every 2 open positions
- Data Scientists – 1 expert for every 4 open positions

Project Successes and Failures

Various sources report that 65-100% of Big Data Analytics projects fail. They are:
- Incomplete
- Out of time
- Over budget

Sources reporting less failure than 65% include POCs, POTs, POVs and pilots in this estimate. Sources range from large infrastructure vendors to Software-as-a-Service providers and open source alliances.

To improve this rate, some sources include a variety of technologies and consider their experience with them to include systems that existed before the addition of Hadoop and NoSQL. The first large vendor ecosystems for cloud and infrastructure premiered in 2011-2013. Platform enhancements such as rules optimizations were the focus in 2007-2010, when some corporations and startups piloted Big Data Analytics open source components in their systems.

Because of this, the classic predictor of success, previous experience, may not be present until new staff and/or vendors are brought in. Experience is required in the current Big Data Analytics technologies, not just in the client's industry or in project management in general. Vendors and teams that say, "We know project management and DWBI, so we can do this without Big Data Analytics experience," are denying themselves the learning they need to be successful.

Why Projects Have Failed

Big Data Analytics projects have been incomplete, out of time and over budget for many reasons that relate to technology and project expertise.

Staffing and Stakeholders

- Could not hire required staff early enough or at all in spite of interviewing many candidates
- Could not find or bring in an experienced vendor early enough or at all
- Did not have sufficient sponsorship and buy-in from the internal or client team or stakeholders

Design

- Did not include the complete ecosystem in the design and requirements development
- Did not have clear data architecture, governance, validation and quality assurance
- Did not anticipate current data needs and new data handling
- Did not identify sufficient components and interfaces during design
- Did not implement an object canon or create any missing custom objects
- Did not plan for appropriate security
- Phase 0 deliverables were not sufficiently reusable to support Phase 1 development

Planning

- Did not scale some of the application components to the build phase
- Took too long to approve QA test accounts for credit cards and mobiles
- Took too long to contract for live traffic testing through communications service providers
- Did not plan for the time required to automate regression testing and prepare simulators
- Budget was pulled before results could be proven
- Phase 1 infrastructure was not purchased, cleared completely through international processes, or installed in the datacenter
- Did not have time to document for support and enhancement work in production

Execution
- Did not discover critical path needs across multiple teams
- Did not implement a central dictionary
- Did not complete application process workflow configurations in time for scheduled testing
- Analytics could not be implemented
- Did not establish clear network virtualization requirements or create VM templates
- Did not implement centralized logging
- Did not complete production preparation such as externalizing logging levels
- Did not load balance and test performance during development, continuous integration and system testing
- Did not automate configuration pushes and application upgrades
- Did not implement network scanning or documentation sufficiently for production
- Did not attach monitoring feeds to the network operations center and security operations center for real-time visualizations and reports

Client Expectation Management
- Did not negotiate metrics, KPIs and reporting
- Did not take time to gather and implement client requirements in a reasonable timeframe
- Did not specify additional localities for data regulations and local terminology
- Did not test host-to-client secure VPNs before production cutover

This list of the reasons for some project failures is indicative, not complete. Note that the classic reasons for project failure can also occur.

Driving for Project Success

There are many technical nuances involved in a Big Data Analytics project, so project managers must be wise enough to encourage exploration and share their learning. We must capture and prioritize discovered requirements, and we must work with the client in the loop under the expectation of an exploratory experience.

Many project management skills are transferable and useful in Big Data Analytics projects for experts and non-experts. A project manager attempting Big Data Analytics for the first time must watch for many factors and issues that can increase risk and cause projects to be incomplete, late and over budget. Projects can be successful if properly steered and nurtured.

Thriving Projects and Project Managers

In any project management and team member role, you have a responsibility to yourself to learn from and teach others. As a project manager, your responsibility is to lead thriving teams and projects. As a program, portfolio or PMO manager, your responsibility is to make this possible for project managers.

Coaching project managers to work with Big Data Analytics technologies and transformational change is the most effective approach. Training is extremely helpful, and peer coaching among project managers and others in project management roles produces the best results.

Project teams often integrate internal and client employees. Simply balancing the needs of the client and the needs of employees is not enough. A productive integration of people delivers successful projects and rewarding work.

Project managers must understand and help the client. Collaborate with the client in order to elicit needs and elaborate innovations. This generates opportunities within projects and for additional projects.

As you proceed with the client and the project team, work together and learn together. Enrich the team with good relationships and interesting, rewarding work.

Keep making progress to deliver a rewarding scope of work and business value. Resolve issues as they arise. Proact whenever possible. Keep your eyes open and your feet nimble. Above all, keep your mind, communication and relationships engaged.

Remember to appreciate people's work and ideas.

Remember work-life balance for everyone, including clients. You may be offered a late-night meeting with the client, but should you accept or wait for a more lively meeting during business hours? Clients must be awake and focused.

Likewise, you and your team are not at your best when you appear run-down with overwork. When client stakeholders ask how the project is going, share one or two focused activities that are delivering success and business value instead of speaking too casually or exhorting the intellectual runner's high from so much activity.

Maslow's hierarchy of needs can become a factor with fast-moving teams. Remember food and sleep. Altoids and ramen noodles are not food groups. Sleep is not that 20 minutes under the desk until there's another issue. In fact, if you build a better system, you will not have to go through this with your teams. Quality matters most. There is no harm in catering lunch or dinner in, or in getting to a stopping point in the work and going home for the day.

Do not burn out team members by engaging the same people to work around the clock for days or weeks. The work is of poorer quality, and it takes longer to code and test due to errors and poor memory of what was done.

PMBOK and Big Data Analytics

The Project Management Body of Knowledge (PMBOK) is a wonderful resource, but it does not address Big Data Analytics. Is this a concern?

As we adopt more and more sophisticated, predictive systems and ecosystem technologies, the quality of our concerns must rise. Our thinking must change, impacting our ways of working. Our processes must advance to keep pace with our predictive analyses and decision insights.

Some areas of PMBOK are impacted by the need for expertise when Big Data Analytics projects are involved. Many considerations are somewhat different from the essential guidance, as PMBOK is not intended to be technology-specific.

On a Big Data Analytics project, there is less delimitation of project, program and portfolio management as team members seek knowledge from anyone who knows more and has the ability to help remove barriers. Escalations may be messy. Timelines are definitely shorter.

Be flexible and do not restrict communication in an effort to reach clarity. Developing knowledge follows a tribal path until ideas can sync up through collaboration. Micro-management and excessive fear about control can over-extend the schedule by as much as 30% due to the additional time needed to resolve issues and take action.

The Big Data Analytics project lifecycle is shorter because it makes more sense to start small and gain experience. We review the project lifecycle and related considerations in depth in the following chapter. Organizational process assets, project management processes and project management wisdom are all impacted by the special considerations of Big Data Analytics.

Stakeholders and organizational influences require a multi-touchpoint approach. This is a team effort of contact with stakeholders at different levels in the client organization. Effective, well-coordinated project management, executive influences and technical grassroots-building can make progress toward more business with the client. This process is different because its domain is a new, hyped technology prompting myriad client concerns and the need for transformative change management.

Do we need a special methodology, BDA-PM?

We expect to extend our thinking to change our ways of working, implying that formal processes and best practices will develop with our wisdom. We need specific expertise and understanding. Because many DWBI experts moving into Big Data Analytics have little or no Agile methodology experience, BDA-PM starts helping them now.

5

Project Lifecycles

Typical Big Data Analytics Project Lifecycle

The Big Data Analytics project lifecycle can be bumpy because it is highly iterative and highly exploratory. Team members may be involved for many sprints and production weeks due to the need for specialized expertise, testing and support.

As projects continue, this can create the impression of some lack of planning, communication, awareness and control. Project managers must set client and internal stakeholders' and the project team's expectations for flexibility.

It is wise to initiate, plan, execute and close in short phases for a series of smaller cycles. Use models, POCs and pilots to minimize risk and explore options easily. Plan to work in parallel by overlapping activities such as development and testing.

Exploit synergies and efficiencies of scale. Bundle code for reuse and redeployment. Use Coding Standards to advance code maturity.

Environments can also be reused and redeployed to create other environments across development, QA, staging, pilot, pre-production and production.

Good design mitigates many risks. Define testing needs early to avoid delays of a few weeks to months depending on the internal and external services needed. Design for easy application process and infrastructure monitoring so that automation can be leveraged and performance can be analyzed predictively.

Governance is needed. Escalations may not solve problems. Don't find yourself and your team running so fast that only critically-escalated work is completed.

Collaborate to solve problems, but don't design by negotiation. Know when the team needs facilitation or simply sleep.

Organizational Process Assets

Necessary organizational process assets (OPAs) do not exist in a complete, clear form at the start of most Big Data Analytics projects because the technologies are new to most enterprises.

Many traditional EDW and BI assumptions are no longer valid. A basic DWBI background and OPAs are not enough to meet Big Data Analytics needs. Templates may be inapplicable or need adaptation.

Estimations of the effort to produce OPAs may miss the target just like estimates of system capacity and development and testing efforts. Prepare for discovery.

Big Data Analytics projects and technologies bring new knowledge and new OPAs, including:
- Coding standards
- Best practices
- Code libraries
- Reusable frameworks, tools and scripts
- Technical data package
- Knowledge transfer
- Training and coaching
- Documentation of design and implementation
- Runbooks and other support documentation
- Knowledge base development
- Terminology definitions and use

Privacy and security constraints can impact OPAs. The delimitation of tasks, access and visibility may make it more difficult to bridge areas of specific expertise.

Big Data Analytics systems need coding for data selection and masking, encrypted databases and security testing of internal code and externally-facing systems. A project may require permission requests with formal approvals to view OPAs, and in-code documentation may be prohibited.

Project Management Considerations

For Big Data Analytics projects, some types of processes are better than others. Traditional process weight may be too heavy for Big Data Analytics iterations. Processes must be lightweight and highly facilitative. CMMI is very helpful.

It may be difficult to analyze process fit for such a highly exploratory effort. Remember that Big Data Analytics must be built as a complete ecosystem in order to function well. The right expertise and collaboration are more important than a finely tailored process in an enterprise's first projects.

To get started with Big Data Analytics projects, use the roadmap, project plan and high-level WBS to identify missing pieces. Then manage with milestones and make progress toward them. Keep everyone in the loop as the activity progresses.

Agile Scrum is the preferred methodology. Others include different Agile methodologies and hybrids such as test-integrated, customer-driven and requirements-based options.

After the first few projects, begin to identify good process facilitations and process pain points so that project management processes can be optimized as projects mature and project mangers gain experience.

Estimation

For a Big Data Analytics project, work estimates have some value but may be off considerably due to the exploratory nature of the work. Waiting for estimates does not tend to make them more accurate even though learning improves estimates. If abstraction is needed, use a modified Fibonacci sequence for Agile software projects such as Rapid Delphi, but do not spend too much time on it. The easiest approach is to create a reasonable roadmap with milestones and then estimate time sprint by sprint unless more planning is needed.

Big Data Analytics code is written in familiar languages, making effort estimations easier, but coding and testing estimates may not take critical factors into account. Prepare for discovery.

Estimations of system volumes, capacity, speed and load balancing require expertise because Big Data Analytics systems process data differently. Standard formula assumptions of partitioning and Big Data-oriented sharding are not enough.

Coders in particular have trouble estimating system capacity for applications because they work with small files and expect only a few files to remain when data processing is complete. If your estimate of the production system with nominal and spiked data flows is based on one-time flow-through of small files, you have not considered the full system functionality and requirements such as all-or-nothing transactions, automatic query-results replay tools and many other services that are part of the complete platform and its ability to deliver 5-9s performance. You must also consider real human behavior such as erroneous queries and sending all queries into the system at the same time. While a coder's granular estimate gives a small average file size, the real system load estimate may show a much larger average file size. Improved system designs can reduce this estimate based on requirements and selected technologies. Smaller file size does not necessarily improve speed if the design is not optimized.

Scope

Scope flexibility may be limited because a complete ecosystem must be built. There are many components in the systems along with operating requirements and related factors. Scope must be monitored carefully.

Application extensions are important, including search and other services. Users must like the product and find it useful.

Automated testing must be coded and used. Performance testing, monitoring and network automation along with backup and failover testing are required. Regulatory compliance must be in place. One security incident can sink the product.

Schedule

Schedule flexibility may be limited due to many external and internal impacts. It is usually best to hold to the schedule and adjust the scope of the release if possible, but this strategy can simply move overwork from development into production support with many manual tasks. Remember that product delivery is success in production, not just in the release handoff.

Beware of the most time-consuming activities that require the most lead time for external entities:
- New vendor contracts
- Infrastructure acquisition, shipping, installation and verification
- External testing services

Quality

Quality is easier to create and sustain on a Big Data Analytics system. Expertise is required to identify requirements.

Data validation ensures that data is correct and can be single-sourced for analytics purposes. Data reconciliation and deduplication eliminate unneeded copies of data records in the system.

All-or-nothing transactions ensure 5-9s accuracy and allow triggers for automatic replay tools to resend queries and results if they do not complete as expected. ACID and BASE models can be blended to provide the best data handling for the system.

Data privacy and security must be enforced throughout the system. Dedicated nodes may be needed to handle encryption-decryption and data gateways. A variety of security best practices for Big Data Analytics provide reliable data privacy.

Staffing Shortages

Staffing shortages are a reality on any project, but the special expertise of Big Data Analytics makes the problem more acute. The client must be willing to pay more for qualified people.

Big Data Analytics experts are hard to find.
  − Hadoop Project and Program Managers −
    1 expert for every 4 open positions
  − Hadoop Developers and Administrators −
    1 expert for every 2 open positions
  − Data Scientists −
    1 expert for every 4 open positions

If no one with expertise is available for hire or through a vendor contract, make the hiring package more competitive. In the meantime, define tasks on an expertise basis and push lower-level tasks to learners so that each project can build team members to have more expertise. Don't be afraid to hire people with transferable skills and train them for more.

Communication

Communication must be collaborative and interactive even if there is a formal definition of roles, responsibilities and interfaces on the project.

As project manager, remember to be a good citizen and communicator. Do not over-ask for estimates and status updates. Provide positive energy, calm and appreciation instead of stress. Your work with others must be a good experience for them.

Problem-solving on a Big Data Analytics project may follow a path of rapid tribal knowledge before ideas can sync up. Do not assume that escalation occurred appropriately as you interact with multiple stakeholders and team members. Be open with everyone so that they can stay in sync.

Big Data Analytics expertise produces a variety of efficiencies. Most problems are real and need solutions. It is critical to take team members seriously when they ask for help. Do not assume that someone who never asks for anything is just now starting and should be declined at first, or you may alienate a highly productive team member.

Note that some projects, such as Sarbanes-Oxley compliance and law enforcement systems, require segregation of duties so that a black box deliverable can pass from development to testing and auditing without bias or influence.

Risk Management and Mitigation

Substantial expertise in Big Data Analytics is required to effectively mitigate and manage risk. Risks can be estimated as part of planning and then updated as changes impact the project.

When performing the risk analysis for a Big Data Analytics project, be more specific than the traditional top 3 risks known to every project plan. As project manager, ask yourself some questions:

- What could happen if the design encounters difficulty?
- What is the next line of knowledge for a team member who cannot get help with an issue?
- How can support engineers collaborate directly with developers and QA testers for a complete requirements loop?
- What are our contingency plans if incoming resources and equipment do not arrive?
- How often are we reprioritizing the PBI in sync with the roadmap and upcoming milestones?
- How will we communicate with executives if a critical escalation is needed and we must proceed without client input?
- What is our contingency plan if there is a security breach during development?

Do your best to spell out risks and how they can be mitigated. This is an important part of planning and ensuring project success.

6

Methodologies

Common Misconceptions

Like any hyped new technology, Big Data Analytics have misconceptions in the market.

Clients are often concerned that "Big" implies a monolith project. Start small and grow. Large systems develop over time. Capacity management planning is easier with Big Data Analytics.

Waterfall methodology is not required simply because the client's business problem qualifies to be solved by a Big Data Analytics system. Your client who loves waterfall can still see regular reporting without your development teams suffering through the methodology, risks and other implications. Agile Scrum works best.

Few clients are actually "dinosaurs" mired in older DWBI. Every conservative, old-style company has hot spots of innovation and employees who keep current in the latest technologies.

Hype does not mean that more burn-in is needed. Hadoop has been in use in enterprises since 2005, picked up steam in 2007 and has expanded exponentially each year since then.

Big Data processing or storage alone does not bring the business value. Do not let clients start with just data processing. Analytics are required.

Analytics without data architecture, governance, validation and quality do not bring the business value. Do not just jump in without a plan for structuring and integrating the data.

Overcoming Client Obstacles

Typical client obstacles to the implementation of Big Data Analytics range from organizational to analytical.

In the current state of the client data warehouse, if one is in use, plenty of data may have been submitted, but little or no analytics action may have been taken.

In the desired state with Big Data Analytics, data does not wait. Real-time triggers are submitted. For example, analytics insights may prompt a real-time offer engine to propose purchases to the customer. The resulting purchase is called the Next Best Action.

Organizational obstacles exist for the client when programs are "owned" by different groups and/or third parties. Different priorities and varying approaches may cause confusion as well.

Data integration can be more difficult when data resides in silos and there is no 360-degree view of the customer.

Analytical obstacles involve the lack of maturity in the client's analytics programs and models. For example, there may be a lack of channel preference segmentation and propensity scores by channel and program.

Usually there is no learning system for refinement of analytics models and algorithms. Change management requires an organizational discipline to embed analytically-driven processes into unified frameworks for interaction with the customer.

Using an incremental approach can reduce gaps. For analytics-driven customer interaction, begin with a subset of channels, programs and members.

Support process design with a learning system as analytics mature. Include a framework for scaling as the system grows.

With solid planning and client expectation management, these obstacles can be overcome.

Do Big Data Analytics fit the client?

The client's business problem must qualify to be solved by a Big Data Analytics system. Begin by reviewing the formula BV = VVCV+SQVV+VVVV. Then ask some more questions as follows.

Which levels of analytics are in place?

Storage Only means that the client is unprepared to begin using analytics and needs a great deal of help to get ready.

Basic Analytics for reporting what happened can provide a solid grounding for further analytics development.

Predictive Analytics for forecasting and proaction put the client in good shape to begin a serious project for advanced predictive analytics and decision insights.

How is data management implemented?

No formal data governance is a great risk. Some data architecture and governance must be in place in order to gain business value from the Big Data Analytics system.

Centralized or decentralized data management or MDM provides enough structure to begin.

What environments are in play?
If Big Data Analytics is part of the client's environment, a hybrid scenario already exists. The client can proceed with more Big Data Analytics but may need help to pull a cohesive strategy together.

If the client wants the cloud, plan to implement Big Data Analytics using one of the project models for the cloud on-premise behind the client's firewall or on a service provider's infrastructure.

The client requiring on-premise technology can use any of the project models and may grow to implement its own private cloud. Infrastructure is required, a major addition to the project.

The client may use only SaaS providers and have no EDW or data science. If the client wants more of this for Big Data Analytics, there are many service providers for data processing, storage and analytics services. A client with no EDW or infrastructure may prefer to implement this hosted model instead of building out large datacenters.

Is client readiness for Big Data Analytics in place?
If the client has no metrics for assessing business value, these must be developed before any meaningful Big Data Analytics can proceed.

If the client has no transformational change management in place, use known best practices.

If the client has no in-house technical or other expertise, supplement with vendor applications, services and staffing until the client can take on complete management and development of the Big Data Analytics system if desired.

Choosing the Best Methodology

Agile methodologies are most appropriate for Big Data Analytics. Many flavors can be effective depending on client situations and the nature of the project.

Scrum works best because of the iterative discovery process. Choose Scrum if there is no reason to choose another methodology. Scrum is familiar to most people and easy to learn fast.

RAD is helpful when global teams work 24x7 and customer requirements need prioritization and implementation immediately. The greatest concern with RAD is its lack of documentation when applied to a new technology that needs OPAs and process analysis to reach maturity.

Customer-driven methodologies involve the client directly with engineering teams. This can work well if the client relationship is good and all participants are able to work meaningfully. The client participants should not be observers only.

Stanford Advanced Project Management is a research-based model that formalizes many aspects of project management wisdom. Its insights and tools are applicable to Big Data Analytics projects. Project managers must use their judgment to evaluate their projects.

There are many Agile methodologies and best practices. Choose the methodology that works best for you and your team based on your existing best practices, processes and knowledge. You can always adapt the methodology as needed.

Waterfall and hybrid methodologies are less applicable to Big Data Analytics. Test-integrated methodologies work reasonably well because they can be more iterative.

It's tempting to try to gather all requirements in advance. V-Model and similar methodologies make an iterative process of requirements development but can wander off track from the critical path if allowed. These methodologies tend to extend the time to completion and do not always build efficiently in exploratory projects.

Scrum and Scrum Again

Scrum is the preferred methodology for Big Data Analytics projects. Even project models that include infrastructure buildout and significant analytics development can benefit from Scrum.

Teams are key to Scrum. Use several Scrum teams for a Big Data Analytics project.
- Development
- Unit and integration testing
- QA, automation and performance testing
- Infrastructure buildout
- Application scripting
- Application configuration for workflows, dictionary, etc.
- Network virtualization and monitoring
- Support

Use Scrum to leverage deliverables and knowledge as they are built. This is especially good for the iterative, exploratory nature of Big Data Analytics projects.

Coordination is critical. This is the best advice:
- Have a daily Scrum of Scrums and report results to management. Without this cross-team coordination, many critical needs may go undiscovered.
- For 24x7 global teams, have a handoff Scrum so that the work can move forward seamlessly.
- For the same team running double shifts, Scrum morning and evening.
- For infrastructure buildout, Scrum morning and evening.
- Audit completions daily and update the burndown chart.
- Do not ask for additional updates!

Evaluating Caveats and Risks

Big Data Analytics projects can be risky if proper risk analysis and management are glossed over to reassure stakeholders. Ask questions!

Start with a small environment.
- Do a Proof of Concept, Proof of Value or pilot.
- Use cloud services (AWS, Cloudera, etc.).
- Save money by building out the pilot to pre-production and production.

Start with a well-defined core dataset.
- What data is available?
- What analytics questions are the most interesting?
- How will we add new data types?
- How will we add new analytics?

Define well, but you do not need to know everything in advance. Think about:
- Processes and workflows
- External integration
- Component decomposition and messaging
- Necessary services

Build by layer.
- Big Data takes in VVCV data.
- Data processes quickly.
- Analytics queries output meaningful insights.
- Data is stored efficiently.

Project Models

While Big Data Analytics may seem like uncharted territory, many models exist to make explanations easier for clients and project managers.

Big Data Hub

The Big Data Hub pulls in a data feeds, integrates the data and makes it available for real-time analytics. Data feeds and analytics results can be passed along to an external entity, another data hub or another analytics platform.

The benefits of this model include high speed, fast integration, efficient handling of disparate and difficult data types, real-time analytics and fast data flow-through to other information recipients.

Caveats can involve unrecoverable data loss if application process monitoring and data protection are not in place. Visualization applications make the data more user-friendly as it flows through the hub.

Big Data Factory

The Big Data Factory provides data processing enhanced with the efficient application of complex verification, validation, governance and quality rules and external services. This model usually includes the infrastructure hardware buildout as well as the applications, virtualization, application process monitoring, network monitoring and related functionality. Any analytics and visualization are focused on keeping the data processing platform up and running with no data loss or security risk.

Benefits include fast data intake, integration and storage.

Caveats can include a longer query-results time and user perceptions of centralization. An analytics platform must be present to derive business value from the data and analytics insights.

Big Analytics Factory

The Big Analytics Factory is an analytics platform enhanced by complex validation, governance, data modeling and quality rules and external services. Infrastructure may be part of the buildout, but typically this is already in place and the factory is simply an application.

Full-insight predictive analytics can be used on data stored in a variety of servers. Visualizations range from simple to complex and cover data interpretation as well as platform operations.

Benefits include efficient real-time and near-real-time analytics and storage of queries and results for future reference.

Caveats can include concern about centralization, the integration of multiple analytics applications, and the integration of data sources from a variety of servers and EDWs. Without a Data Factory or Data Lake, query-results times can be longer.

Big Data Foundry

The Big Data Foundry is an optimized Enterprise Data Warehouse made more interesting by the efficient intake of real-time and near-real-time data. Today's Foundries include analytics applications, visualization tools, SaaS and IaaS.

The benefits of the modern Foundry include fast processing and storage plus optimized querying based on rules and complex data architecture.

Caveats can include the perception that a Foundry is old technology with isolated data, the ability to handle large data volumes in a clustered environment, the agility of the system in general, and the perception that a large install base is required.

Big Data Lakes

Big Data Lakes are designed to make data fluid and accessible given real data volumes and user requirements. Lakes are often separated logically as real-time temporary processing and analytics, storage processing and query-results, specialized data processing, and specialized analytics. The

Lake can be anywhere in the environment if that is the most efficient way for the rest of the system to deal with the data and its operating requirements.

Benefits include greater agility when adding commodity gear, easier enterprise-wide access to data for analytics, and clear delimitation of functionality and ownerships.

Caveats can include concern about over-decentralization, the variety of applications accessing the Lakes, and the more complex elegance of design granularities.

Big Data Analytics Center of Excellence

The Big Data Analytics Center of Excellence can be implemented internally and for clients. The CoE has three streams of activity: business engagement, governance and knowledge.

The major business value of a CoE is its support of business engagement by evaluating Big Data Analytics use cases to qualify them for use with the technologies, estimate their effort and prioritize them, burning them down one by one to enable business and technology units. Advisory services are an important part of this effort.

Data governance on the basis of clear data architecture is required to gain value from Big Data Analytics. This ensures enterprise governance to enable a unified Big Data Analytics strategy and optimize research and development activities. Established data governance provides platform governance for future capabilities.

Because Big Data Analytics form a new technology with a complex ecosystem, deliberate knowledge-building and knowledge-sharing is needed. Along with the CoE, a knowledge base includes client and vendor product demonstrations to inspire thinking and evolve strategies. Client and internal stakeholder workshops extend beyond demos and strategy to review new technologies, new use cases, results from POCs and POVs, lessons learned and deliverables. Stakeholder awareness grows to interactive conversations about successes, reuse, re-engineering and evolving needs.

Caveats can include concerns about centralization and decentralization, appropriate sponsorship and budget, and lack of knowledge across the enterprise. It is critical to have a friendly sponsoring business unit with more BUs willing to participate in use case development, advisory services and education activities.

Big Data Analytics Delivery PMO

For a client who provides Platforms-as-a-Service and on-premise solutions to enterprise customers, a specialized Big Data Analytics Delivery PMO can be established as a subunit of the enterprise-wide

PMO and as an infusion team within the Big Data Analytics business and technical practice. The PMO delivers rapid, repeatable implementations for enterprise customers.

The benefits of a specialized PMO include the new technology learning for the existing PMO and the grounding in substantial project management expertise for the new teams implementing Big Data Analytics.

Caveats can involve many of the topics addressed in this book.

Big Data-as-a-Service

Big Data-as-a-Service can take many forms: social media hubs, industry-specific business and technical applications, Infrastructure-as-a-Service, monitoring solutions and more. These solutions can involve dedicated nodes, custom data ingestion, advanced semantics and custom workflows, dictionary and related items.

The benefits involve cost reduction and ease of automation, customization, deployment and application value management (enhancement, maintenance and support).

Caveats can include the shortage of expertise and the need to adhere to strict SLAs for enterprise customers.

Big Analytics-as-a-Service

Big Analytics-as-a-Service makes the analytics platform available in the cloud to internal and external entities. Client products and partner platform-sharing fit this model.

This is a low-cost way to offer customized, advanced analytics for enterprise customers who know the benefits of cloud services or who cannot build their own platforms or data science capabilities in any other way. Good analytics increase revenues and enable good decision-making with visualization tools.

Caveats can include concern about customizing analytics for particular business needs as well as concern about data privacy and security.

4-Week Business Case Development

In lieu of an assessment, a business case development project can focus and define the business value and the use case for Proof of Concept testing and later implementation.

This project has a flexible structure and is designed to complete quickly with solid value using the following steps, activities and deliverables.

1. Readiness Review – Activity: Client Preparation. Deliverable: Prerequisites and Readiness Review Questionnaire. Note that this step may not be client-billable.

2. Use Case Selection – Activity: Use Case Matrix. Deliverables: Use Case Wireframe and Industry-Specific UC Repositories.
3. Use Case Definition – Activity: Requirements. Deliverables: Value accelerators.
4. Business Case Development – Activity: Update Business Case Template. Deliverable: Business Case Wireframe and Guidelines.
5. Business Case Review – Activity: Business Case Reviewed with Business Sponsor.
6. Go or No-Go Decision.

The biggest benefit is rapid demonstration of business value to begin a fast Proof of Concept.

Caveats can include lack of specificity and clarity.

2-Week Proof of Concept

Following the business case development, the POC can be accomplished in about 2 weeks. A clear scope of use cases and business complexity for the POC is required to begin.

If the client has a long list of use cases, select the best. Do not include more than 3 use cases.

This project has a rapid-results structure and provides clear business value using the following steps, activities and deliverables.

1. (As Above) Go Decision on Business Case.
2. Lab Setup – Activity: Set up the POC environment. Deliverable: Big Data Labs. Note that the Big Data labs can be reused with subsequent POCs, POVs and POTs.
3. Use Case Execution – Activity: Build the use case application. Deliverables: Value accelerators.
4. Value Demonstration – Activity: Demonstrate the business value. Deliverables: Value accelerators.
5. Go or No-Go Decision. If Go:
6. Implementation – Activity: Develop and deploy. Deliverables: Value accelerators.

The biggest benefit, again, is rapid demonstration of business value.

Caveats can include completing too quickly or not quickly enough for the defined scope and expectations.

The Business Use Case + POC cycle can repeat through as many use cases as the client wishes to explore. The Big Data Analytics Center of Excellence uses this model to deliver business engagement across business units. This model is effective for any new use case implementations if their complexity requires a POC.

12-Week Strategy Assessment

A strategy assessment is a longer-term project involving interaction with many stakeholders. Often the strategy assessment precedes a project model implementation such as a platform or CoE.

Activities

- Introductory Workshop, Week 0
- Engagement Planning and Kick Off, Weeks 0-1
- Project Management, Weeks 0-12
- Big Data Strategy Assessment, Weeks 1-12
    - Business Strategy and Priorities Discovery, Week 1
    - Business Information Needs Assessment, Weeks 1-3
    - Business Use Case Development, Weeks 2-5
    - Conceptual Architecture, Vision for the Cloud, Technology Guidance, Weeks 3-9
    - Big Data Analytics Center of Excellence Definition and Strategy, Weeks 6-11
    - Big Data Analytics Governance and Strategy, Weeks 6-11
    - Transformational Change Management, Training and Communication Plan, Weeks 9-12
    - Roadmap and Implementation Plan, Weeks 10-12

The purpose is to elicit and review the current strategy, explore needs and determine how to drive to the desired business value and state of technology, data architecture, governance and quality.

A defined engagement model should be used that includes team members from business and technology groups so that a good influence pattern can engage the client stakeholders.

24-Week POC + Implementation

Oftentimes the goal of the POC is to win the implementation project by demonstrating that real business value will be delivered for the client.

For this long-term project, do not plan in detail beyond 24 weeks. Planning is not effective beyond this point due to the high level of exploratory development and data changes.

Have a clear objective and be sure that the activities and staff can support and reach the objective.

The Proof of Concept phase, like other POCs, is short and flexible to allow for exploration.

Proof of Concept Phase

Up to 4 weeks

Sample Objective: Develop one next-generation analytics and visualization view.

Activities
- Scope definition/validation
- Select data types
- Obtain sample data
- Analyze and correlate data
- Render real-time data flow
- Segment data
- Provide a Rollout baseline and plan
- Develop high-level roadmap (draft)
- Develop high-level business plan
- Gap analysis of current and desired analytics and business value

Deliverables
- POC results
- Roadmap
- Draft high-level business case

When the POC has completed successfully, the project can proceed with more POCs if needed to explore the platform concept, or the initial plan for implementation of the POC's use cases can begin.

Note that POC environments can be leveraged to build development, QA, staging, pre-production and production environments.

Analysis and Rollout Planning Phase
4 to 8 weeks

This phase concerns the readiness of the client to adopt the POC's use cases in a specified way.

Activities
- Stakeholder workshops
- Define business objectives based on POC results
- Detailed review of current state processes
- Detailed review of current state technology (mediation, data storage, analytics, network interfaces, data interfaces)
- Detailed review of current state reporting metrics/KPIs
- Detailed gap analysis and mitigation plan
- Define future state roadmap and tactical plan
- Develop draft rollout plan
- Obtain sample data from each source
- Project expected results for each query type

Deliverables
- Current state maturity mapping
- Gap analysis and results
- Future state roadmap and tactical plan
- Draft rollout plan

Once readiness and planning have been established, proceed to implementation.

Implementation Phase

2 to 12 weeks

Activities
- Roll out solution to all entities
- Benchmark each entity
- Prepare markets (People)
- Prepare business (Process)
- Prepare infrastructure (Technology)
- Communications plan
- Align roadmap with relevant initiatives
- Identify key projects and prioritize by BV/ROI
- Develop baseline rollout plan
- Develop and present results
- Identify next steps for additional data types and business units

Deliverables
- BU-by-BU rollout
- Business plan benchmarking against expected results
- Executive results summary

When the implementation is complete, sustain business value through the best practices of application value management for enhancements, maintenance and support.

Along with a great Big Data Analytics system, be sure to provide complete service packages to clients, including application and infrastructure support in multi-year managed services contracts.

Be a trusted advisor and partner to the client. This ensures more opportunities to help the client in the future. Successful implementation is not the last step in the client relationship.

7

Unique Considerations

Platforms and Policies

Big Data Analytics require a special data processing platform with unique considerations for project managers and teams.

Data Governance

Data architecture must take into account the specific needs of Big Data processing. Only with proper data architecture can we build the foundation of data governance, validation and quality upon which insightful, meaningful analytics can thrive and extend.

Degrees of centralization and decentralization must be present in enterprise-level Big Data Analytics strategies. Business unit proximity provides valuable insights into use cases.

As new BU use cases are proposed, they must be evaluated and prioritized before they can become projects.

Metadata must be managed well. It is possible to externalize metadata management and processing as a service plugging into the Big Data Analytics platform.

For content systems, metadata is the key to identifying and delivering the correct content that a user has requested. Metadata processing and use in searches can be audited for performance reporting to stakeholders.

Reconciliation and deduplication allow the fewest number of copies of data to be retained and leveraged in the system. Any changes to data values must be fed through the system and updated everywhere until processing is complete.

Data governance and retention policies must deal with data at its server location. Transitory data and backup, disaster recovery and site-to-site failover must be architected, implemented and tested in order to ensure 5-9s reliability and resilience.

Security and Privacy

Big Data Analytics systems have complex needs.
- Encryption is required on all components.
- Data masking and tie-to IDs must be used whenever data might be visible, even in logs.
- Server locations may impose data policies.
- Data retention must adhere to regulations.
- Customer data requests must be allowed.

Services Worth Having

- Email verification ensures that a real person with a functioning email address is interacting with the system.
- IP verification ensures that a legitimate node is interacting with the system.
- GPS data lookup provides latitude and longitude coordinates that can plot a device's location on a map.
- IP location lookup estimates the GPS of the IP address in use based on previous activity.
- Audit history makes it easy to prove good performance and debug issues.
- Externalized metadata speeds up busy systems with metadata centralized outside the platform as a plug-in service.

Application Process Monitoring

To ensure that transactions, queries and results complete, application process monitoring watches for non-completions, paused queries, empty folders, acknowledgment failures, missed logging and other indications that a process has not completed properly.

Network Monitoring and Automation

To ensure 5-9s reliability and resilience, network monitoring and automated issue resolution relieve teams of manual tasks. Reporting can flow to NOC and SOC visualizations while alerts trigger automatic memory provisioning and route traffic.

Test Automation

Regression testing can be automated for most Big Data Analytics systems using proprietary or open source applications and tools. Performance test automation allows traffic variations to test for nominal and spiky traffic levels for hours or days. For mobile and platform test automation, a combination of mobile applets and server scripts is good for uptime verification pings, performance burn-in, GPS and related testing.

Some required manual testing is always needed for credit cards, mobiles, locations and other factors. Testers can focus on more complex tests with a high level of user interaction.

Compliance Reporting

Compliance reporting takes two forms, backend data pushes and standard reporting/visualization.

If data must be reported to a compliance agency, it can be pushed by rules or pulled by querying over the backend of the system. This is efficient for security, fraud prevention, eDiscovery, SEC, FDA and rules-based compliance.

Reporting and visualization is also needed for the same data in the form of counts. For example, today 10 queries for eDiscovery were processed. We don't see the topic or individuals involved. Counts and durations make good operating proof.

System Metrics and KPIs

Big Data Analytics systems may require special metrics/KPIs due to the nature of data processing in the system. Counts and durations are good measures of application performance in addition to standard network performance measures. Focus metrics/KPIs on data intake, processing, queries, results, process completions and automatic fixes triggered in the system.

See Chapter 8 for Big Data Analytics project metrics/KPIs.

Growth and Sustainability

How can Big Data Analytics systems adapt, streamline and grow?

New data types and workflows arrive weekly in high-traffic and social media-oriented systems. New data type handling and related semantic technologies are critical resources for Big Data Analytics systems. This need will increase until technology convergence is achieved.

Cluster management and node forecasting must be present to allow systems to automatically provision themselves as needed.

Automated deployment, network and application process monitoring is required for performance purposes. Replay, backup, recovery and failover must also be automated.

Cloud technologies are the natural home for Big Data Analytics systems and accumulating user preference, UX and historical data storage.

Social media and mobile integration will become seamless with Big Data Analytics systems.

Software-Defined Networking (SDN) is software that discovers needs and provisions the system. This is clearly headed for convergence with Big Data Analytics technologies as we redefine and reimagine high-performance computing.

How can we frame our thinking to optimize business value?
We must ask ourselves some key questions.
- What are we trying to accomplish? How is this valuable?
- What data questions do we have now? What questions may follow?
- Can the opportunity or problem benefit from Big Data Analytics?
- Does the solution have growing room?

We must design for the future as well as for today's needs and technologies.

8

Program, Portfolio and PMO Management

Big Data Analytics poses challenges to project managers in many ways, including impacts on program, portfolio and PMO management. The project manager who understands these dynamics internally and among client stakeholders has a better chance of project success.

Contexts, Impacts and Dynamics

Big Data Analytics requires new thinking and even new ways of working. Expertise is essential. Dealing with hype is only the beginning.

External Impacts

Market hype has made plenty of high-level information available about Big Data Analytics, but which expectations are substantial and appropriate for the project at hand? Information sources can vary widely in their opinions and research about technologies, staffing needs, business value and return on investment. Amid the noise, expectations must be set with a degree of exploratory processes and incremental rewards. This requires expertise in Big Data Analytics.

Stakeholders are often cautious. Is it the right time for Big Data Analytics? Does the business problem qualify to be solved with this new technology? What has been overlooked? Who will be held accountable for success or failure?

On a Big Data Analytics project, the project manager must prove that the methodology is the right one for the project. Agile is generally favored, but some clients demand hybrid-waterfall variations. Most client technical employees have heard of Agile Scrum even if business stakeholders have not. A PMO that is well-versed in Scrum may have questions about its execution given the highly exploratory nature of the Big Data Analytics work and the narrowly-defined expertise requiring many roles to collaborate in order to accomplish the project. Using the same Scrum may not be enough.

Under these circumstances, a clear scope of the project, the roadmap, each release, each sprint and each story is essential at the right time. Story prioritization criteria must be used objectively. Some flexible time must be available to resolve unanticipated problems and allow for the presence of previously undefined and new data types in the system. The team cannot deliver the business value without clear scope and flexibility.

Delivering the business value on a Big Data Analytics project means more than achieving results as expected before the project began. Some clients want a measurable return on investment. Most prefer incremental ROI and demonstration of business value once they know it is possible. A few believe hype that tells them not to focus on ROI because they won't see it for 2 years after production in any case, which is untrue. The client can see business value and ROI as soon as it is clearly defined in a small enough scope to be implemented.

In dealing with stakeholders and team members who interact with them, the project manager must position for growth. Big Data Analytics mean more volume, variety, complexity and velocity along with more analytics insights, business value and revenues if the benefits of the technologies are realized. Understanding customers, business conditions, technologies and enterprises provides a competitive edge.

- We grow because we get it.
- We grow more intelligently than our competition.
- We grow in spite of internal resource competition and politics.

When using this fast-growing, hyped-up technology, it is important to balance elements of centralization and decentralization among projects, workstreams, programs, portfolios and the PMO. Big Data Analytics is game-changing. It is not enough to allow expertise to dictate a silo approach. The PMO that isolates itself or allows one facet of its efforts to isolate may find itself with barbarians at its gate.

To provide real, long-lasting business value, leverage Big Data Analytics as the reliable, fast-growing technology that it is. Let others play. Use the new technology and its path toward ubiquity to raise the water level and with it all of the boats in the company.

It is not enough to speak as a thought leader and trusted advisor to clients. A deeper collaboration and growth must happen. Set up processes and tools that facilitate this. Let people and their ideas matter.

Growth creates the need for outside help from vendors of technologies, solutions, consulting, services and staffing. Because Big Data Analytics technologies are new and hyped up, they attract new players attempting to gain work with limited experience, expertise and staff. Be sure to carefully evaluate the qualifications of vendors and allow the participation of those with transferable experience and skills. For example, HDFS is the Hadoop database. CDFS is Cloudera's HDFS. They are substantially similar, so transferable experience and skills, and the willingness to learn more, may be enough for your project. It is not necessary to find someone who has done exactly this type of project or worked with this product before. If you decide to wait for perfection, you may wait longer than most of your competitors.

Many vendors offer solutions that pair hardware and cloud software. They sell or resell traditional stacks that have been optimized for Big Data Analytics, partnering with well-reputed providers such as AWS, Cloudera, Hortonworks and MapR.

Hadoop data processing can be delivered as a service in the cloud, typically through SaaS or PaaS. Clients prefer a private cloud with secured data. Hardware can be located on customer premise or in vendor datacenters. Some clients use Big Data Analytics in their cloud products and provide Big Data-as-a-Service and Big Analytics-as-a-Service to their enterprise customers.

With so many new things to talk about, vendor presentations can be very high-level, with every new face wanting to educate about Big Data Analytics. Be sure to clearly articulate with the vendor what you expect for the pitch. If you want education, say so. If you've heard it already, identify some use cases for a deeper discussion with the vendor. What do you want to accomplish? Go ahead and tell the vendors. You may attract mostly vendors who can provide what you need.

In your own presentations to your clients, remember to get the intelligence first by having some casual conversations with client stakeholders. Then present your capabilities. If you find your clients interrupting your presentation frequently to ask questions, you may not be feeding them enough new information fast enough. If they sit quietly and do not interact with you, you may not be eliciting enough information from them.

As Big Data Analytics has gained strength in the market, its usefulness has expanded the possible client stakeholders from the traditional IT groups and business authorizers to individuals in business, marketing, sales and even security. To deal with them effectively, gather some intelligence so you can demonstrate that you understand their perspectives.

IT groups today are pressured to know the most about Big Data Analytics while having limited exposure if they are trying it for the first time. They need a lot of help but may not have a lot of budget. Be careful to socialize the idea of paying for projects such as POCs, or you may wind up having invested poorly.

Business authorizers, if you meet them at all, want to see clear business value in people's behaviors and documentation as readily as the recommending stakeholders can. Be sure to attend to the details and provide a version of your capabilities and project that gives a business authorizer confidence in you. It's wise to show cost savings as part of the proposal.

Marketing stakeholders may have some or no technical acumen. They may need basic education and reassurance in Big Data Analytics technologies, ease of use and business value. Their teams may include data scientists as well as marketers.

Sales stakeholders are driven toward business value but may have limited technical acumen. Make sense fast about business value, especially if your solution is complex.

PMO stakeholders need to see that you can deliver business value and control costs by running projects well. You know yourself. Be prepared to show that you can achieve your goals, learn from experience and transfer knowledge to the client employees.

Security stakeholders can be strange birds if you have not dealt with them before. Do not be surprised if the CSO must approve the project. Be prepared to show how you apply security best practices and policies. Cover any laws and regulations related to your project, especially if privacy, fraud prevention and compliance are concerned. If you do not know the answer to a question, say that you will find out and make sure that you do this.

The C-suite stakeholders, if you are fortunate enough to speak at this level, are interested in business value and elegant simplicity. Hype does not interest them, but a serious overview of Big Data Analytics without hype may be a winner. This advances the relationship even if a project is not immediately available.

The sales cycle for Big Data Analytics can be short or long. This is important from a planning perspective for program, portfolio and PMO management. Be sure to track the progress of projects through a pipeline from opportunities to realized revenue.

Some clients have been desperate to get started for a long time but were not able to find an appropriate vendor or staff with the right skills. A few clients may seem fickle for a few months and then return to mount a serious initiative. Other clients have not started to think through what they might do with Big Data Analytics. This means that the sales cycle may vary from a couple of months to a year or more.

Internal Impacts

External impacts produce internal impacts specific to Big Data Analytics considerations and dynamics.

Because Big Data Analytics is new, there is a lack of predictive analytics about these fast-moving projects. Clients may not have basic analytics because they have not attempted any projects. Vendors and even you may lack meaningful analytics for the same reasons.

Begin with the reliable success/failure metrics and KPIs used on most projects. Drive learning and quality based on your Big Data Analytics expertise.

Effective Measures
- On time, now and projected
- Within scope given flexibility, now and projected
- Within budget, now and projected
- Initiation of items and preparation with enough lead time (hardware acquisition, testing services, automation tasks, etc.)
- New data types and their stories added to PBI
- QA, UAT and change management completions
- POC, POV and POT completions
- Reusable code and framework completions
- Application process monitoring completions
- System expansions (hardware installations, VM propagation, monitoring metrics, etc.)
- Knowledge transfer to production support

Less Effective Measures
  - Methodology adherence is unlikely to be the biggest concern with the new technology.
  - Process compliance may demonstrate a lot of discovered deviations and stops as learning progresses.
  - Count of team members is inflated due to narrow expertise and partial-FTE assignments.
  - Lines of code and sizes of configuration files produced are not that valuable given system architecture, operations and automation.
  - Successes/failures show a lot of partial successes based on starting small and learning.

Among multiple projects, there are synergies of expertise that allow narrowly-defined experts to serve in multiple efforts.

As reusable components are produced among projects, they can be adapted into other projects. Significantly complex components can become accelerators and even productized as a platform.

A Big Data Analytics project manager must be able to serve as a methodology guru and coach for team members. These gurus and coaches must rely on program managers, portfolio managers and the PMO to develop their own knowledge and coaching. Be sure to encourage peer learning and knowledge-sharing among project managers.

Tension can develop around choosing the right methodology for a particular project and team due to the newness of Big Data Analytics and how much an end-to-end platform encompasses.

Because the technology is new, PMs need access to technology wisdom as well as project management wisdom. A lot of project issues are resolved when their technical aspects are addressed.

Facilitative Structure

PMO services, portfolios, programs and projects must be structured to facilitate the needs of Big Data Analytics projects, which are exploratory and have less established knowledge for baselining.

Managing requests and escalations can be difficult. At first, it may seem like every request or issue is completely new.

A thriving PMO may take on leadership for the technical practice, especially in some project models. Typical requests and escalations include:
  – Technology, strategy and business value
  – Help with clients, conferences and capabilities
  – Use cases, POCs, POVs, POTs and pilots
  – Special questions and consultations
  – Help with processes, reporting and metrics/KPIs
  – Staffing, training and coaching
  – SLAs with clients and internal business units
  – Vendor proposals and interest

Managing the project hierarchy requires an expert sense of how projects can synergize on the basis of teams, clients and technologies. Projects can become more time-consuming as they progress. Eventually a repeatable cycle will emerge that stabilizes project sizing estimates. Remember to assign PMs and project staff with knowledge-building and cross-pollination in mind.

Managing methodologies involves expertise in the technologies and the selection of an appropriate methodology for each project and client. Forcing a poorly-fitted methodology onto a Big Data Analytics project can cause enough confusion to derail its potential success. It is more important to use the best methodology than to have all projects in the PMO, program group or portfolio following the same methodology. Gain experience first and then adapt toward the standard if needed.

Any methodology used must be Agile or hybrid rather than waterfall in spite of the temptation to determine all requirements in advance. Big Data Analytics is a large ecosystem with many players and many ways of accomplishing the same result with different tools, models, coding and approach. Every project is at least somewhat exploratory from kickoff into production support. A solution design process helps evaluate architecture options before the resulting solution starts development. Coders should have some flexibility in the design.

Managing documentation requires more effort than usual due to the newness of Big Data Analytics. Documentation may not exist or may not be created easily because nothing similar has been done before. Any common templates may need customization before use. Projects move quickly and do not allow a lot of time to document details. Coding standards may prohibit in-XML documentation due to security policies or the lack of processing to separate and journal those notes. Even knowledge base features such as keywords and tagging involve a new set of terminology and a technically relevant logical structure of folders and documents. It is best to start small with what is critically needed and mature quickly as the project progresses.

Adaptation and Leadership

Organizational adaptation strategy requires a balance of enterprise response to the external environment and leadership. Organizational adaptation is any process of adjustment within the enterprise that improves its equilibrium in the market, its customer bases, its supply chains and its distribution networks, if not its entire community.

This is different from the internal growth and development of an enterprise and its employees. Organizational development and related topics such as organizational change management and corporate training use internal strategies.

Adaptive Balance

Big Data Analytics can transform the nature of the enterprise and how it works. The style of environment response and the importance of leadership during change produce four types of strategy orientation, each with its impact on project, program, portfolio and PMO managers working with Big Data Analytics.

Population Ecology is survival of the fittest, with or without leadership. This can create confusion, tribal knowledge and other conditions that require a project manager, who can be very effective in rallying teams to accomplish tasks with high uncertainty.

Life Cycles produce a sequence of development phases similar to incremental models in technology. Phases often repeat in the same order, but this is not required for each phase to be effective. Each phase needs project and program management.
   – Creativity and entrepreneurship
   – Collectivity and sophistication
   – Formalization, control and efficiency
   – Elaboration of structure to fuel decentralization and expansion

Strategic Choices rely on situational actors and events to baseline adaptation. Specific information, locations and moments of choice produce the actions of prospectors, analyzers, defenders and reactors in the marketplace. Project managers must be dialed into their client and internal environments with the help of program managers. Information gathering is required to leverage advancements.

Symbolic Action is more theoretical and defined as the social construction of reality and roles. Enterprises venturing into social media must take on these skills in their online presence and marketing campaigns. Savvy project and program managers are needed to put the enterprise voice into the market and hear other voices. The skill is also useful in internal and client political environments and stakeholder networks.

It is important to remember that people follow these four strategic models rationally and instinctively. Whichever model is prominent, the other three can sometimes be seen bubbling under the surface with various stakeholders, which can indicate a shift from the current strategic orientation to another one.

Organizational Leadership

Enterprise transformation requires both organizational adaptation and transformational leadership. Big Data Analytics provide information that can make many new ideas possible. Project, program, portfolio and PMO leaders must consider this internally and with clients.

Leadership during change tends to emphasize similar frameworks for providing structure, focusing on people, encouraging teamwork and pointing toward values and goals. These ideas focus on the leader driving the enterprise culture and business activities.

In a thriving enterprise, leadership must be practiced in different contexts by different people. In general, an enterprise with a high level of innovation and collaboration is poised to get the most from its people and give them the most rewarding work experience. Having too few people can be a serious barrier to this advancement. An internal political landscape of multiple isolated stakeholders with competing factions and alliances can slow growth and adaptation in spite of the quality and appropriateness of leadership and who practices it.

Experienced consultants know that this all boils down to internal and client engagements. For Big Data Analytics projects, a more transformative context for activity is needed to deepen relationships and drive future business activity.

Big Data Analytics projects can be won with granular employees as well as in the C-suite.

Especially in technology, a groundswell of thought leadership can make significant progress. It can be interesting enough for everyone to want to play together in spite of silos and competition. With better understanding of the productive balance of centralization and decentralization of data and technologies, teams can keep what they value most while improving the way in which it works.

Project, program, portfolio and PMO managers must act with leadership to emphasize this possibility and help make it reality.

Reflective Leadership and constant communication are considered good strategies. In practice, the reflective leader may seem reticent and repetitive, doing little while saying the same things to the same people. Leaders who are too comfortable with change being driven externally may find themselves appearing disinterested to their teams.

Action-oriented leaders display the passion and energy that is infectious and builds confidence. To be effective, this must accompany substantive actions, responsiveness and progress.

Collective Leadership as an interactive practice involves key elements. Leadership occurs in different contexts by different people. Effective leadership includes the skill of taking the perspective of the other people as part of diplomacy, discussion and decision-making.

Leadership is a relationship, not a transaction. Business and employment are inherently transactional to create a thriving economy and provide everyone with resources and choices, but this is not the basis for a relationship. There are many employers, teams and vendors.

People choose to work well together. A relational context such as a hierarchy or functional role does not really create or define a relationship. It is best if we believe that all of us are inherently worthy and have good contributions instead of taking a less mature approach to being in charge or the most expert among the others.

Transformational Leadership leverages change management and interactive communication strategies. In isolated and internally competing enterprises, this can produce strategic realignment and strategic planning. Marginal transformations can guide behavior for a time, but a new marginal transformation may be needed soon. If culture is shared in a highly collaborative enterprise, everyone's leadership can pull together to create highly productive, effective, synergistic collaboration among business units, teams and projects.

Organizational Learning requires a high level of collaboration, coaching, knowledge management and access to advancement for employees. Risk-taking must be a way of life for the enterprise because it must teach itself how to learn from itself. Big Data Analytics are a great driver for this, especially when we begin to leverage predictive analytics. Projects must consider these questions:

- Do we know what we know?
- Do we know what we don't know?
- How do we know if we know or not?
- How can we learn from others?
- How can we teach ourselves?

Ambidextrous Organizations communicate well enough to know about activities across the enterprise and with the client. The right hand knows what the left hand is doing.

Janusian Organizations conceptualize and organize along the lines of style, strategy and experience in responding to certain types of external impacts and internal dynamics. One enterprise has multiple "faces" to show based on positioning the needed advancements internally and externally in the market. Leaders must be careful not to tell too many different value and capability stories to the same people. Projects fit into this model well, but it can be hard to unite factions in order to advance the business beyond project-by-project work.

Big Future Organizations know themselves well enough to respond to external impacts and internal dynamics with behavior patterns that revitalize and renew the enterprise. Big Data Analytics are a large part of this capability.

Discrete strategic orientations such as Life Cycles and Strategic Choices become activity frameworks supported by the most effective leadership from a variety of involved people. Core values, shared leadership, passionate innovation and shared culture drive the action. The enterprise is agile and driven with solid grounding and unique ideas.

Predictive analytics and enterprise-wide access to their insights are required to achieve this level of enterprise capability and effectiveness. Project, program, portfolio and PMO managers can make exponential positive impact internally and with clients when they use these principles and explain them to others.

Hands-On Exercise

For more fun with Big Data Analytics, try this planning exercise.
- Create a repeatable model for a client's Big Data Analytics in a focus area that inspires you and your team:
  - Industry-specific system
  - Operations and support systems
  - Networking and infrastructure
  - Applications and integration
  - Research projects
  - Investments
  - Regulatory compliance
  - Adaptive language and distance communications
  - People and social life
  - Marketing and advertising
  - Education, employment and recruitment
- Continue this exercise and discussion via social media and online applications.

Remember to download and play with the open source tools and demo applications that are available online from various sources.

References

<u>Search online</u> using the terms in this book.

For good information in one place, visit these links.

Hadoop
http://hadoop.apache.org/

NoSQL
http://nosql-database.org/

AWS Ecosystem
http://aws.amazon.com/

Data Science
http://www.datasciencecentral.com/

Agile
http://agilemanifesto.org/

Scrum
http://www.scrumalliance.org/

NOTES

NOTES

Made in the USA
San Bernardino, CA
13 November 2019